Luftwaffe Air Crews
Battle of Britain
1940

by Brian L. Davis

Published by
Arms and Armour Press
Lionel and Leventhal Limited
2-6 Hampstead High Street
London NW3 1PR

SBN 85368 107 4

Series Editor: Brian L. Davis, for
Key Military Publications
Series Design: David Gibbons, for
Arms and Armour Press

Typesetting: ABCDesign
Camerawork: Duotech Graphics
Limited
Printed in England

Acknowledgements

The author wishes to thank Martin
Windrow for his invaluable
assistance and the Department of
Printed Books of the Imperial War
Museum, London, for access to
basic Luftwaffe reference material.
Photographs reproduced are from
the collection of the Imperial War
Museum, and Bundesarchiv,
Koblenz (plates 11,18,20,23,25,
26,28,32,34,39,40,42,43,44,45,48,
49).

Plate 1, below: *ME109 maintenance crew welcome comrades returning from a sortie in the late summer of 1940.*

Contents

Historical Background

The Luftwaffe

Officially constituted in the spring of 1935, the Luftwaffe was built up into an instrument of German national policy of a most impressive size and its existence was undoubtably one of the chief causes of the fear which spread and deepened over Europe from 1935 on. Yet, in fact, the German Air Force was never as powerful as it appeared to be and it eventually proved totally inadequate for the task it was called upon to do. In the early 'blitzkrieg' years of the war it gave good tactical support to the German Army, it pressed the Royal Air Force to the limit of its endurance in the Battle of Britain, but eventually, as the war progressed into its middle years and the tide of battle turned against the Germans it was slowly but inexorably bled to death.

The prewar development of the Luftwaffe and its organisational structure at the time of the Battle of Britain

The organisational structure of the Luftwaffe reflected the function for which it was developed — the tactical support of large and mobile ground forces, as opposed to the defence of the Reich and its territories, or the launching of strategic attacks from fixed home bases. This policy was the direct result of the change in emphasis brought about in 1936 with the death of Generalleutnant Wever. Wever was, in his time, the one man in the German Air Ministry whose foresight and ablility to plan and co-ordinate could have changed the face and fortune of the Luftwaffe during its formative years, and no doubt during the initial and very crucial years of the war. As the Luftwaffe's first Chief of the Air Staff, he had laid long-term plans for the development of the new German Air Force. Included in these plans was the use of large numbers of four-engined heavy bombers. Wever saw the long-range bomber as the means whereby Germany, as a central power, could exert her will by striking out at any corner of Europe. Had these plans for a strategic bomber force materialised Germany, equipped with long-range heavy bombers, would by 1939 have been almost invincible and probably capable of delivering a crushing weight of bombs on the United Kingdom before the end of 1940. These plans, however, were destined never to mature, for on 3rd June 1936 Wever was killed in an air crash near Dresden. The plans for Germany's heavy bomber fleet died with him.

Albert Kesselring succeeded to the post of Luftwaffe Chief of Staff and, with Göring, he proceeded to examine the bombers then under development. For a variety of reasons, which included shortage of materials, the imminent prospect of land warfare with neighbouring countries as well as Generaloberst Ernst Udet's[1] preference for fast evasive tactics, it was decided to concentrate on fighters (Messerschmitt 109s and 110s), dive-bombers (Junkers 87s, known as 'Stukas' from the German 'Sturzkampflugzeug', meaning dive-bomber) and medium bombers (Junkers 88s, Dornier 17s and Heinkel 111s). All this was at the expense of the heavy four-engined bombers, in which the Luftwaffe remained notably deficient throughout the war.

The tactical role of the Luftwaffe required corresponding mobility on the part of the associated air formations and this in turn led to the development of an organisational system in the traditional German Army pattern — the smaller units established on a functional basis with a single type of aircraft, and the higher headquarters commanding a mixture of units and types.

The basic Luftwaffe flying unit, roughly comparable to the Royal Air Force squadron but slightly smaller, was the 'Staffel'.

2. Two members of a bomber crew pose for an aerial portrait. Both are wearing the unlined tan-coloured flying helmet and the crewman on the right is also wearing Nitsche & Gunther shatterproof flying goggles.

[1] Generaloberst Udet was at this time in charge of Luftwaffe technical development.

3. Above: *German fighters sweep the English Channel.*

This comprised ten to a dozen aircraft of the same type — fighters, dive-bombers, or others as the case might be.[1] The Staffel often flew as a formation of ten aircraft, with three 'Ketten' of three aircraft each normally led by the 'Staffelkapitän'. Three Staffeln comprised a 'Gruppe' which approximated a British 'wing' and at full establishment would consist of 39 aircraft (12 in each of three Staffeln and an additional 'Kette' constituting the staff ('Stab') aircraft of the 'Gruppenkommander'. In turn three Gruppen constituted a 'Geschwader'— literally a squadron— the largest homogeneous aircraft unit of the Luftwaffe.[2] Normally the Geschwader had 120 aircraft (39 planes in each of the three Gruppen, and the staff Kette of the 'Geschwaderkommodore') but, owing to the decline in numbers brought about through

[1] Other aircraft formations were: a 'Rotte', a tactical formation of two aircraft, the basic fighter formation, normally made up of a pilot officer with his 'wing man' (Rottenflieger), usually an NCO; and a 'Schwarm', a formation of four aircraft — two Rotte.

[2] Staffeln were denoted by arabic numerals, while Gruppen had Roman numerals, e.g: II/JG51 (meaning Second Gruppe of Jagdgeschwader 51); 2/JG51 (meaning Second Staffel of Jagdgeschwader 51).

combat conditions, there were rarely as many as a hundred, and during the course of the Battle of Britain there were often as few as eighty planes to a Geschwader.[1]

Gruppen and even Staffeln could and did operate independently from widely separated airfields, but during the Battle of Britain they were closely concentrated. Administratively the Geschwader headquarters remained the parent unit even if its components were scattered, except for reconnaissance aircraft and a few other types that were organised in Gruppen but not in Geschwader.

Each Geschwader bore the designation of the type of aircraft of which it was composed. The level-flight bombers, mostly twin-engined Heinkel 111s and Dornier 17s were in 'Kampfgeschwader' (KG—Bomber squadrons); the Junkers 87 dive-bombers formed 'Sturzkampfgeschwader' (StG — Dive-bomber squadrons); the single-engined fighters, virtually all Messerschmitt 109s, comprised the 'Jagdgeschwader' (JG — Fighter squadrons); the twin-engined Messerschmitt 110s were called 'Destroyers' and their squadrons accordingly 'Zerstorergeschwader' (ZG — Destroyer squadrons). There were also 'Transportgeschwader' (Transport squadrons) of three-engined Junkers 52s and special units of other types, such as night-fighter squadrons — 'Nachtjagdgeschwader' — and 'Lehrgeschwader' (LG — Training squadrons).[2]

From the Geschwader the ascending chain of command normally passed up to the 'Fliegerkorps' which was intended to include 'Geschwader' of different types and therefore to be able to carry out the various missions of the air arm. During the Battle of Britain there were six 'Fliegerkorps' (numbered I, II, IV, V, VIII and IX) on the Channel coast and one (X) stationed in Scandinavia. Although the Fliegerkorps was supposed to be a 'general' headquarters, in practice some of them became specialised: for example, Fliegerkorps VIII, commanded by General Wolfram Freiherr von Richthofen, was almost exclusively composed of 'Stukageschwader' and Fliegerkorps IX specialised in minelaying and other aspects of marine aviation.[3]

Above the Fliegerkorps were the Luftwaffe's highest field headquarters the 'Luftflotten' (Air Fleets). In peacetime there had been four of these; a fifth was added in Norway in April 1940, during the German conquest of that country.[4]

Whilst the Battle of Britain was being fought Luftflotte 1, commanded first by General der Flieger Wimmer (acting commander) then by Generaloberst Keller, and Luftflotte 4, commanded by General der Flieger Lohr, remained at their home stations (Berlin

---

[1] Variation in the number of aircraft making up different types of formations was due to battle losses, replacement availability but above all serviceability of aircraft. Servicing accounted for units normally being 20% below their physical establishment.

[2] Many of the Geschwader had honorary designations: JG 2 was Jagdgeschwader 'Richthofen', JG 26 was Jagdgeschwader 'Schlageter'; KG 1 was Geschwader 'Hindenburg'; KG 53 was 'Legion Condor'; ZG 26 was Geschwader 'Horst Wessel'; StG 2 was Geschwader 'Immelmann'.

[3] These Fliegerkorps were originally called 'Fliegerdivisionen' and at the start of the Battle of Britain the IX was still called 'Fliegerdivision 9'; it was upgraded to 'Fliegerkorps IX' in October 1940. There were no Fliegerkorps numbered III, VI, or VII. Fliegerdivision 7 comprised the Fallschirmjäger and Luftland forces (Parachute and air-landing forces); in 1941 Fliegerdivision 7 was upgraded to become 'Fliegerkorps XI'.

[4] The peacetime headquarters of the four prewar Luftflotten were: Luftflotte 1, Berlin, covering north and east Germany, commanded by Kesselring; Luftflotte 2, Brunswick, covering north-west Germany, commanded by Felmy; Luftflotte 3, Munich, covering south-west Germany, commanded by Sperrle; Luftflotte 4, Vienna, covering south-east Germany, Austria, and the occupied parts of Czechoslovakia, commanded by Lohr Luftflotten 1, 2 and 3 were created in February 1939, Luftflotte 4 in March 1939.

and Vienna respectively) virtually stripped of their aircraft. The Battle of Britain was waged by Luftflotte 2, commanded by Generalfeldmarschall Kesselring, based at Brussels and Luftflotte 3, commanded by Generalfeldmarschall Sperrle and based at Paris. To a much lesser degree Luftflotte 5 was employed in the battle and this was commanded by Generaloberst Hans Jurgen Stumpff, based in Norway.

The Luftwaffe Order of Battle, 13th August 1940: Battle of Britain

Luftflotte 2

Headquarters Brussels, Belgium. Commanded by Generalfeldmarschall Albert Kesselring. Bases in Holland, Belgium and Northern France. Component units were:
Fliegerkorps I: Headquarters Beauvais, France. Commanded by Generaloberst Ulrich Grauert. Consisted of:
Kampfgeschwader 1
Kampfgeschwader 76
5 Staffel, Aufklärungsgruppe 122
4 Staffel, Aufklärungsgruppe 123
Fliegerkorps II: Headquarters Ghent, Belgium, Commanded by General der Flieger Bruno Lörzer Consisted of:
Kampfgeschwader 2
Kampfgeschwader 3
Kampfgeschwader 53
II Gruppe, Sturzkampfgeschwader 1
IV (Stuka) Gruppe, Lehrgeschwader 1
Erprobungsgruppe 210
II Gruppe, Lehrgeschwader 2
Fliegerdivision 9: Headquarters situated between Amsterdam and the Dutch coast at Soesterberg, Holland. Commanded by Generalleutnant Joachim Coeler. Consisted of:
Kampfgeschwader 4
Kampfgeschwader 40
Kampfgruppe 100
Kampfgruppe 126
Küstenfliegergruppe 106
3 Staffel, Aufklärungsgruppe 122.
Jagdfliegerführer 2: Headquarters Wissant, North-East France Commanded by Generalmajor Theodor Osterkamp. Fighter and Destroyer forces in North-East France. Consisted of:
Jagdgeschwader 3
Jagdgeschwader 26
Jagdgeschwader 51
Jagdgeschwader 52
Jagdgeschwader 54
I Gruppe, Lehrgeschwader 2
Zerstörergeschwader 26
Zerstörergeschwader 76.

Luftflotte 3.

Headquarters St.Denis, Paris. Commanded by Generalfeldmarschall Hugo Sperrle. Bases in North and North-West France. Component units were:
Fliegerkorps IV: Rear Headquarters at Compeigne, Forward Headquarters at Dinard. Commanded by Generaloberst Kurt Pflugbeil. Consisted of:
Lehrgeschwader 1
Kampfgeschwader 27
Sturzkampfgeschwader 3

4. Top left: *Generalfeldmarschall Albert Kesselring, Commander of Luftflotte 2, August 1940.*
5. Top right: *General der Flieger Wolfram Freiherr von Richthofen, Commander of Fliegerkorps VIII, August 1940.*
6. Below left: *Generaloberst Ernst Udet, Director of the Technical Dept. of the German Air Ministry, 1936-1939, and Director General of Air Force Equipment (Generalluftzeugmeister) from 1939 to 1941.*
7. Below right: *Generalfeldmarschall Erhard Milch, appointed Secretary of State for Air (Stactssekretar der Luftfahrte) 26 February 1935.*

8. *General feldmarschall Hugo Sperrle, Commander of Luftflotte 3, August 1940.*

Kampfgruppe 806
3 Staffel, Aufklärungsgruppe 31.
Fliegerkorps V· Headquarters, Villacoublay, France. Commanded by General der Flieger Robert Ritter von Greim. Consisted of:
Kampfgeschwader 51
Kampfgeschwader 54
Kampfgeschwader 55
Fliegerkorps VIII: Headquarters at Deauville, France. Commanded by General der Flieger Dipl.-Ing. Wolfram Freiherr von Richthofen . Consisted of:
Sturzkampfgeschwader 1
Sturzkampfgeschwader 2
Sturzkampfgeschwader 77
V (Zerstörer) Gruppe, Lehrgeschwader 1
II Gruppe, Lehrgeschwader 2
2 Staffel, Aufklärungsgruppe II
2 Staffel, Aufklärungsgruppe 123.
Jagdfliegerführer 3: Headquarters, Cherbourg, Northern France Commanded by Oberst Werner Junck. Fighter and Destroyer forces on Central Channel Front. Consisted of:
Jagdgeschwader 2
Jagdgeschwader 27
Jagdgeschwader 53
Zerstörergeschwader 2.

Luftflotte 5
Headquarters Stavanger, Norway. Commanded by General-oberst Hans-Jurgen Stumpff. Bases in Norway and Denmark. Component units were:
Fliegerkorps X: Headquarters at Stavanger. Commanded by Generalleutnant Hans Geisler. Consisted of:
Kampfgeschwader 26
Kampfgeschwader 30
Zerstörergeschwader 76
Jadgeschwader 77
Küstenfliegergruppe 506
1 Staffel, Aufklärungsgruppe 120
1 Staffel, Aufklärungsgruppe 121
Aufklärungsgruppe Oberbefehlshaber der Luftwaffe
Aufklärungsgruppe 22.

Battle of Britain Battle Phases
Phase One (10 July to 7 August 1940)
German air attacks on Allied shipping and British coastal ports. Aerial tactics employed by German fighters prove superior to British fighter tactics. The Boulton-Paul Defiant I, the turreted fighter, proves useless against German fighters. Whereas fighter production is no longer a serious problem for the RAF there is a very serious lack of trained pilots. The British concentrate on raising pilot strength and building up for the battle ahead: Air Chief Marshal Dowding draws heavily on pilots from RAF Coastal Command and the Fleet Air Arm. Four Polish and one Czechoslovak squadrons formed within a matter of weeks.
Losses Phase One: Royal Air Force Fighter Command: 169. Luftwaffe 192 + 77 damaged.

Phase Two (8 to 23 August 1940)
German attacks on British radar stations and forward RAF fighter bases. 13 August 1940 known as 'Adlertag' — Eagle Day. On this day the Luftwaffe deploys 2,442 aircraft against Great Britain in a total of 1,485 sorties: 969 level flight bombers, 336 dive-bombers, 869 Me 109 single-engined fighters and 268 twin

engined Me 110 'Destroyer' fighters ('Göring's Folly'). The climax of this phase occurs on 15 August with the Luftwaffe operating 1,790 sorties. Attacks from Scandinavia (Luftflotte V) are repulsed with heavy losses and after this the Stukas are withdrawn from the battle. On 17 August Germans establish an 'operational area' around the British Isles. In this area all ships are to be sunk without warning. In the south Fighter Command suffers heavy losses and pilots begin to show signs of extreme fatigue. RAF casualties are in the region of 15 to 20 pilots killed or wounded every day. Göring makes a fatal decision and, although he has almost brought Fighter Command to its knees, unwittingly spares the RAF by abandoning attacks on radar stations.
Losses during Phase Two: Fighter Command 303. Luftwaffe 403 + 127 damaged.

Phase Three (24 August to 6 September 1940)
German bomber attacks on aircraft production factories and inland fighter bases. Bombers are accompanied by strong fighter escorts in an attempt to draw up British fighters. On 25 August the RAF conducts its first bomber raids on Berlin. The Messerschmitt Bf 110 and the Stuka dive-bombers have proved very vulnerable to the superior Spitfire and Hurricanes. But British pilot losses and pilot fatigue have reached desperately high levels.
Losses Phase Three: Fighter Command 262. Luftwaffe 378 + 115 damaged.

Phase Four (7 to 30 September 1940)
German attacks on London. In a final effort to destroy British air power after the realisation that Fighter Command is still a force to be reckoned with despite its losses the Luftwaffe switches to daylight bombing raids on London. 7 September sees the first of the heavy daylight 'Blitz' raids on the British Capital. 300 bombers escorted by 600 fighters attempt to destroy London's dockland. Battle reaches its climax on Sunday 15 September. The RAF claims to have shot down 183 German aircraft during Luftwaffe daylight raids on Britain — a figure subsequently found to be greatly exaggerated. On 17 September 'Operation Sea Lion', the proposed invasion of the British Isles, is postponed indefinitely. Germans switch tactics to high-level hit-and-run bomber raids.
Losses during Phase Four: Fighter Command 380. Luftwaffe 435 + 161 damaged.

Phase Five (1 to 31 October 1940)
The aftermath. German fighter-bomber sweeps and preparation for the night Blitz on London. Fighter Command reserves in aircraft and pilots increase rapidly.
Losses during Phase Five: Fighter Command 265. Luftwaffe 325 + 163 damaged.

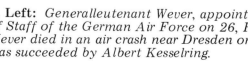

9. Left: *Generalleutnant Wever, appointed as the first Chief of Staff of the German Air Force on 26, February 1935. Wever died in an air crash near Dresden on 3 June 1936 and was succeeded by Albert Kesselring.*

Uniforms

The uniforms, flight clothing and equipment in use by the Luftwaffe Air Crews during the period of aerial warfare that came to be known as the Battle of Britain can be divided into two main groups:
1. The uniforms which, in effect, were the normal issue Luftwaffe clothing complete with headdress and footwear.
2. The Luftwaffe flight clothing which was in use at that time![1]
The purpose of this book is to describe and illustrate these uniforms and flight clothing, including the headdress and footwear and, to a certain extent, the equipment used by German Air Crews. The insignia related to these uniforms and clothing items are dealt with in part only. (All aspects of insignia and uniform for Luftwaffe personnel will be covered in a forthcoming volume).

Luftwaffe Uniform for Air Crews in 1940

In keeping with all officers, non-commissioned officers and men of the German Air Force, the Luftwaffe crews that took part in the aerial assault on the British Isles in 1940 were issued with, or purchased privately (or sometimes both, depending on rank) a full complement of uniform clothing, in addition to the issue flight clothing.

The uniform items listed below are those which the author feels come within the scope of this work — relevant only to flight crews during the Battle of Britain:

The Luftwaffe service tunic (the 'Tuchrock')
The flying service blouse (the 'Fliegerbluse')
The uniform tunic (the 'Waffenrock')
The uniform peaked cap (the 'Schirmmütze')
The flying cap (the 'Fliegermütze')
Luftwaffe trousers and breeches
Luftwaffe shirts
Luftwaffe footwear (non-flight footwear).

Uniform clothing such as the Luftwaffe officer's white summer uniform, the formal full evening dress, the informal full evening dress, the special tunic for Luftwaffe Generals etc have been deliberately omitted from this book.

The Luftwaffe Service Tunic: The 'Tuchrock'

This was a single-breasted tunic which incorporated smartness with practicality. With only slight differences in quality and minor modifications, such as rank insignia, it was worn by all personnel in the German Air Force, including Fallschirmjäger. The basic design for this tunic had been adapted in March 1937 from the service tunic in use by the Deutscher Luftsport Verband (DLV), which in turn had been influenced by the German Army service tunic in use by the Reichsheer. It was, however, intended to replace this Tuchrock, as well as the Fliegerbluse, with a new style of jacket (to be known as the Luftwaffe Waffenrock) in November 1938. This new jacket, which would eventually replace the two earlier models, would incorporate features from both the earlier tunics, whilst at the same time economise on material and labour.

The Tuchrock, as worn by all ranks of the Luftwaffe below that of officer, was blue-grey in colour and manufactured from a wool-rayon mixed material. Luftwaffe Generals, officers and individuals who privately purchased jackets for themselves would

10. Above left: *During a tour of German fighter bases in France Göring visited most Luftwaffe units engaged in the Battle of Britain. Seen here Reichsmarschall Hermann Göring, speaking to Major Adolf Galland is accompanied by (left to right) General der Flieger Loerzer, Oberst Werner Mölders and Generalleutnant Forster.*
11. Below left: *Göring speaking with a pilot Feldwebel during an inspection of a Luftwaffe fighter station in France, 1940.*

---

[1] It should be noted that air crews from Luftwaffe units involved in minelaying and other forms of marine aviation sometimes flew as 'mixed crews'; that is, with German naval aviators flying together with Luftwaffe personnel.

usually wear a fine blue-grey gabardine quality material (plate 12). The tunic had four box-pleated patch pockets. Four metal pebble-surfaced buttons were positioned down the front of the jacket with a similar button to each of the four straight-edged pocket flaps. These buttons were in gold coloured metal for wear by Luftwaffe Generals and Generalfeldmarschalle, and in silver-aluminium finish for all remaining Air Force ranks.

The appropriate Luftwaffe style collar patches as well as shoulder-straps (normally removable) were worn according to the wearer's rank, and on the right breast running parallel to the breast pocket was positioned the Luftwaffe version of the National Emblem (plates 14, 23). For NCOs and men this was machine-embroidered in grey cotton yarn on to a backing cloth of blue-grey material matching that of the tunic. Officers below the rank of General normally wore the emblem in hand-embroidered silver-aluminium wire (plate 23), while Generals and above displayed eagles embroidered in gold wire.[1]

The edging to the collar of this tunic was piped in a 0.3cm twisted cording which for NCOs and men was in the wearer's appropriate Waffenfarbe and for all officers below the rank of General, including Senior Flight Ensigns, in silver-aluminium cording (plate 23). Generals and above wore twisted gold cording.[2] Belt-hook holes, consisting of three reinforced eyelet holes posit-

---

[1] NCOs and men were permitted to wear hand-embroidered National Emblems on privately purchased non-issue tunics.

[2] The pratice of wearing collar piping was discontinued during the spring of 1940 and the order relating to this included those Tuchrock still in use at that time.

12.Left: *Generaloberst Keller, Commander of a Fliegerkorps, wearing the Luftwaffe General officer's Tuchrock.*

13. Below left: *A Luftwaffe lieutenant arriving at a London railway station under guard on his way to an internment camp. This officer wears the Luftwaffe Tuchrock, breeches and heavy fleece-lined flight boots.*

14. Below centre: *Three German airmen shot down in air battles over Britain on their way to a prison camp. The Oberleutnant left is wearing the Luftwaffe Officer's Tuchrock, the Oberleutnant centre is wearing the officer's version of the Luftwaffe Fliegerbluse and the Unteroffizier right is wearing the NCO's Fliegerbluse without the Luftwaffe National Emblem.*

15. Below right: *The Luftwaffe officer's Fliegerbluse.*

ioned one above the other, were situated at waist level on both the left and right side seams of the tunic. Two removable blue-grey metal belt hooks could be positioned at one of three varying heights to support the wearer's leather belt. There was also a small 15cm deep vent in the centre at the rear of the jacket skirt (plate 11). Unlike the German Army service tunic all Luftwaffe service jackets had 16cm deep turn back cuffs regardless of the wearer's rank (plate 11).

The official Luftwaffe regulations prescribed the Tuchrock to be worn with blue shirt and black tie for the following purposes:
1. Parade dress for all Luftwaffe ranks
2. Service dress for officers and Portepee-Unteroffiziere
3. Undress uniform for officers
4. Reporting dress
5. Flying service uniform for all flight personnel in place of the Fliegerbluse.

In addition to this the Tuchrock worn together with a white shirt and black tie was used for:
6. Walking-Out Dress for all Luftwaffe ranks.

The service tunic worn together with a white shirt, a stiff white collar and a black tie were to be used for the:
7. Informal and formal full dress for men and NCOs
8. Informal and formal day time full dress for officers.

The Luftwaffe Flying Service Blouse: The 'Fliegerbluse'
As with the Tuchrock, the Fliegerbluse was issued to all Luftwaffe personnel. Although it also was intended to be replaced by the Luftwaffe Waffenrock introduced in November 1938, it continued to be worn right up to the end of the war.

Its design was something completely new in the Luftwaffe's sartorial style. It was a short, single-breasted, fly-fronted, blue-grey wool/rayon jacket without cuffs to the sleeves; initially the jacket for NCOs and men had no side pockets, while for officers there were no flaps to the side pockets (plates 14, 23). It was designed to be worn open or closed at the neck (plate 14). Originally intended for use by crews of aircraft, the jacket was therefore designed so that, when not worn under issue flight clothing, there were no visible buttons, patch pockets or cuffs to catch on any projecting parts of aircraft interiors. The jacket, however, proved to be sufficiently convenient and smart enough to be worn universally throughout the Luftwaffe, including Engineer and Navigational Corps and Administrative officials as well as by Fallschirmjäger troops as a popular form of dress.

The Luftwaffe rank insignia, as described for the Tuchrock was worn on this blouse. The small pebble-surfaced metal shoulder-strap buttons were in gold coloured metal for Generals and above and in silver-aluminium for all other personnel. One set of two belt-support hooks were worn on the NCO's and men's version of the flight blouse, but none were used on the officer's blouse. The Luftwaffe National Emblem was displayed on the right breast of the officer's Fliegerbluse at all times (plate 17); this was not the case with NCOs and men of the Luftwaffe. Pre-war and up to 1 October 1940 no Luftwaffe National Emblem was displayed on the Fliegerbluse as worn by NCOs and men (plate 14). Its wear was introduced on the 1st of that month. The rule governing the colour and quality of the Luftwaffe-style National Emblem, as described above for wear on the Luftwaffe Tuchrock, also applied to the Fliegerbluse.[1]

The Fliegerbluse was prescribed to be worn with the collar closed or open (in which case it was worn with the regulation neck band) as:
1. Flying service uniform for NCOs and men
2. Field dress for NCOs and men
3. Service dress for NCOs without Portepee and men.
The Flight Blouse was also worn, open at the neck, with the Luftwaffe issue blue-grey shirt and black tie as:
4. Flying service dress and field dress for officers
5. Undress uniform for NCOs on leading duty
6. Undress uniform for officers, in place of the Tuchrock.

The Luftwaffe Uniform Tunic: The 'Waffenrock'
First introduced for wear in November 1938 as a standard article of Luftwaffe clothing, the Waffenrock or uniform tunic was intended to replace both the earlier Tuchrock and the Fliegerbluse. Because of its design and smartness of style this jacket was considered suitable wear either as a peacetime garment or for use on active service. Very similar in appearance to the Tuchrock it differed noticeably in two basic features: firstly, the collar was manufactured in such a way as to enable it to be worn open or closed at the neck, and secondly, there were five buttons on the front of the tunic (as against four on the Tuchrock), all of which were visible when the tunic collar was worn closed at the neck. The collar fastening was achieved by the addition of this extra tunic button on the front of the jacket. The lapels were cut

---

[1] The officer's version of the Fliegerbluse had 'slashed' side pockets with curved flapless pocket openings (plate 15). Other ranks had no pockets to the Fliegerbluse (plates 14, 22). An order dated 19th November 1940 introduced side pockets for newly manufactured flight blouses for wear by NCOs and men (plates 16, 20). Existing old-style pocketless flight blouses were not to be altered but were to continue in wear until replaced by the new issue.

16. Right: *German Airforce wireless operator with the rank of Feldwebel wearing the Fliegerbluse for NCOs and Other Ranks with side pockets.*

17. Below: *A German lieutenant shot down into the sea during the Battle of Britain and rescued by fishermen is brought ashore to receive first aid treatment. This prisoner is wearing the Luftwaffe officer's Fliegerbluse with 'slashed' side pockets, silver corded piping to the collar and the officer's service belt.*

in such a way as to allow them be to closed across the wearer's neck, the left lapel lying under the right collar with the collar hooks fastening the necks together.

In all other features the 'new' Luftwaffe Waffenrock was the same as the 'old' Tuchrock. Both had four box-pleated patch pockets, deep turnback cuffs to the sleeves, white aluminium pebble-surfaced buttons (gilt in the case of Luftwaffe Generals and above) with one button to each pocket flap. Standard Luftwaffe insignia was displayed, including collar patches and shoulder-straps. Twisted piping to the edge of the collar was worn in the appropriate Waffenfarbe for NCOs and men, silver-aluminium for officers and gold-yellow cording for Generals and above; this practice, however, was discontinued in 1940.

The Luftwaffe version of the National Emblem was positioned above the right breast pocket and was manufactured in exactly the same fashion as for the Tuchrock: grey cotton machine-embroidered yarn on blue-grey backing material for NCOs and men, silver-aluminium hand-embroidered wire emblem for officers up to and including the rank of Oberst, and in gilt bullion wire for Generals and above.

The Luftwaffe Blue-Grey Shirt

The issue Luftwaffe blue-grey shirt with long sleeves and attached collar was permitted to be worn by pilots in place of the tunic when on flights during hot summer weather. The sleeves were usually rolled up to the elbow, the collar left open at the neck or worn closed with a black tie (plate 18), and the wearer's shoulder straps fixed to the shoulders of the shirt. The Luftwaffe version of the National Emblem was stitched in position above the right breast shirt pocket. Military decorations were frequently worn on the shirt.

Luftwaffe Waffenfarbe

Reference has been made in this work to 'Waffenfarbe'. This was a system of colours (similar to that used by the German Army) allocated to various branches of the German Air Force by which personnel of these branches displaying their appointed colours on various insignia worn on, and as part of, their military uniforms were recognisable as belonging to a particular rank of the Luftwaffe. The basic colours in use by the Luftwaffe at the time of the Battle of Britain were:

White	Luftwaffe Generals
Black	Construction Troops
Carmine	Luftwaffe General Staff
Gold Yellow	Flight personnel, gound crews and paratroops
Bright Red	Luftwaffe Flak Artillery
Gold Brown	Signals
Dark Blue	Medical personnel
Rose Pink	Luftwaffe Corps of Engineers
Light Green	Aerodrome Supervisory Service Units
Dark Green	Luftwaffe Administration Officials.

Luftwaffe Trousers and Breeches

Long trousers of blue-grey material, matching the colour and texture of the wearer's jacket constituted correct wear for certain styles of uniform dress in the Luftwaffe (plate 14, 20). Officers and certain non-commissioned officers were permitted to wear blue-grey breeches (plate 12, 13, 18, 19).

Luftwaffe Non-Flight Footwear

Footwear in the Luftwaffe consisted of black leather lace-up shoes, officer's high boots in black or dark brown leather, and

black leather marching boots ('Jackboots'). The latter were not normally worn by flight crews, being used only by ground personnel and Luftwaffe ground troops.

The Luftwaffe Schirmmütze

This was the uniform peaked cap, which was worn by all ranks of the Luftwaffe as well as by Luftwaffe administrative officials. It was prescribed to be worn with the service dress and undress uniform by officers and NCOs, with parade dress by officers when not actually taking part in the parade, and with walking-out dress and with dress uniforms by officers. It had (for all ranks) a blue-grey top, a black ribbed mohair material capband (later changed to black artificial silk) and a black shiny peak. Other cap distinctions depended on the wearer's rank and position.

All NCOs and men wore shiny black patent leather chinstraps and black chinstrap buttons to their caps (plate 10), with the Luftwaffe style of the National Emblem and the oakleaf cluster flanked by stylised wings stamped out in white aluminium (plate 20). (The Reichskokarde remained the same colour for all ranks of Luftwaffe personnel with a slight difference for Luftwaffe Generals and Luftwaffe officials of equivalent rank.) Piping used on the NCO's and other ranks' version of the Luftwaffe Schirmmütze was 0.2cm thick, appearing around the crown to the cap and around the top and bottom edges to the black capband, and was in the Waffenfarbe of the wearer.

Luftwaffe officers below the rank of General wore the same style and basic colouring to their uniform caps but they used silver cap cords with small matt silver coloured buttons to hold the cords in position on the cap (plate 21). The cap was piped in silver aluminium 0.2cm thick (plate 26). The peak of the cap was of a shiny black fibre, underside green, with stitching running along the edging to the peak. The insignia worn on the officer's Schirmmütze was normally in hand-embroidered silver wire, the size and design of the badges being the same as those described for NCOs and men. These uniform caps for wear by officers as well as those used by Luftwaffe Generals were usually of a much better and finer quality material than those worn by NCOs and men.

Luftwaffe Generals and Luftwaffe administration officials of equivalent rank wore the same basic style of Luftwaffe Schirmmütze as worn by officers, but in place of the silver cap cords they wore gold coloured cords and small gilt cap cord buttons. All insignia worn on the cap was in gold coloured hand-embroidered bullion wire (plate 12). The uniform cap for Generals and above was piped, regardless of the wearer's branch of service, in gold piping 0.2cm thick, both around the crown and to the top and bottom edge of the black capband. For all ranks of the Luftwaffe the National Emblem was displayed in the centre of the cap front, positioned above the capband and just below the piping around the crown of the cap. The Reichskokarde surrounded by the oakleaf cluster and flanked by stylised wings was positioned on the front of the cap in the centre of the capband directly below the National Emblem (plate 25).[1]

The Luftwaffe Fliegermütze

This fore-and-aft style forage cap was a standard item of issue

---

[1] The Reichskokarde as used by the Luftwaffe differed from the National Colours Rosette used by the German Army and Navy in one small detail. In addition to the red centre, white (or silver) and black outer rings to the cockade the Luftwaffe version of this emblem was edged with a fine white outer ring (plates 22, 23) which in the case of Luftwaffe Generals and Luftwaffe Administration Officials of equivalent rank was gilt.

18. Above left: *At the end of a series of successful sorties against England the health of this German pilot officer is toasted by his comrades. The officer on the right is wearing the Luftwaffe blue-grey shirt and black tie.*
19. Above right: *A German pilot officer wearing his Fliegerbluse tucked into his breeches and high boots being assisted ashore by a British Army sergeant and a Police Constable. Shot down over the South-East coast of England this pilot was rescued from the sea by a British lifeboat.*
20. Below left: *German flight crew non-commissioned officers. All are wearing the Luftwaffe Fliegerbluse and NCO's Schirmmützen.*
21. Below right: *The Luftwaffe officer's Schirmmütze. This headdress was not the normal form of flight headgear.*

22. Above left: *The Luftwaffe Fliegermütze for NCOs and Other Ranks. An Oberfeldwebel taken prisoner during the Battle of Britain passing through a London station on his way to an internment camp somewhere in Britain.*

23. Above right: *The Luftwaffe Fliegermütze for Officers. Hauptmann Weinreich, holder of the Knights Cross to the Iron Cross and the German Cross in Gold.*

headdress to all ranks of the Luftwaffe (plates 20, 42) as well as Luftwaffe administrative officials. Its use was prescribed for those occasions that were not covered by the wearing of the Luftwaffe steel helmet or the Schirmmütze. The blue-grey Fliegermütze was of the same cut for all ranks and services of the Luftwaffe (plate 14), officers usually providing for themselves caps of better quality material.

The Luftwaffe Eagle-and-Swastika National Emblem was positioned on the upper part of the headdress situated in the middle of the vertical seam on the front of the cap, the lower arm of the swastika just reaching the upper edge of the cap's turn-up. Directly below the National Emblem in the centre of the turn-up the Reichskokarde was displayed (plates 22, 23). (Only the cockade, without the oakleaf cluster and wings, was used on this form of headdress). The National Emblem for Luftwaffe NCOs and men was in machine-embroidered matt grey cotton yarn, for officers up to the rank of Oberst in fine silver aluminium hand-embroidered wire, and for Generals and above in hand-embroidered gilt bullion wire. The Reichskokarde worn on this headdress was edged with a thin outer ring which for NCOs and men was in grey cotton yarn (plate 22), for officers in silver wire (plate 23) and for Generals in gilt wire.

Luftwaffe officers, officials of both the Luftwaffe Engineering and Navigational Corps, Luftwaffe administration officials with ranks equivalent to Air Force officers, as well as those non-commissioned officers intitled to wear officer's quality cap cords to their Schirmmütze, wore the Fliegermütze with the upper edge of the turn-up piped in 0.3cm wide silver cord (plates 11, 23, 26, 29, 31, 40). For Generals and administrative officials of General's rank this 0.3cm wide cording was in gilt.

24. Above: *A typical inter-mixing of protective flight clothing. The lieutenant wearing the Luftwaffe summer wear flying suit with the fleece-lined leather flying helmet and leather flying gauntlets.*

Luftwaffe Protective Flight Clothing

The protective flight clothing and flight equipment used by German air crew members during the Battle of Britain was in no way different from that in use during the earlier 'Blitzkrieg' periods or, for that matter, that used in the immediate prewar years. Positive development, with new ideas being introduced into flight clothing and — to a certain extent — with experimental clothing designed to counteract adverse conditions encountered by flight crews, tended to be developed later in the war. Many of these experimental designs failed to reach maturity owing in almost all cases to lack of sufficient raw materials; those that were produced came too late and in too few numbers. The items of flight clothing and equipment listed below are those which were in normal use by German Air Crews during the Battle of Britain.

The Luftwaffe summer flying suit
The heavyweight winter flying suit
The fleece-lined two-piece winter flying suit
The electrically heated flying suit
The Luftwaffe flight jerkins
The lightweight summer flying helmet
The fleece-lined leather flying helmet
The Luftwaffe steel helmet
Fleece-lined flying boots
Gloves, gauntlets and goggles
Inflatable lifejackets
Kapok-filled lifejackets
Sleeping bags.

Flying Suits

At the time of the Battle of Britain the Luftwaffe had three basic patterns of protective flight clothing (flying suits) for use by its air crews with the addition of a number of other forms of flight clothing.

1. *The lightweight tan coloured summer flying suit for flights over all types of terrain.* This was manufactured from heavy-duty cotton material (plates 24, 25, 26). It was a single one-piece 'step-in' garment (see colour illustration inside front cover). It was normally worn with the unlined flying helmet of matching material and with gloves, also of the same coloured material, together with black leather shoes or boots. Rank insignia was worn on both arms of the suit, between the elbow and shoulder seams, by all ranks from Unteroffizier to Generalfeldmarschall (plates 24, 25, 26). The insignia of rank was stitched on to a tan material cloth backing.

2. *The heavyweight fleece-lined flying suit for use during winter months when flying over land.* This fleece-lined suit sometimes referred to as the 'Bulgarian Suit' was manufactured from heavyweight dark blue-gray material. It was a one-piece step-in garment somewhat similar in design to the tan coloured summer flying suit but had a black fleece-covered collar (plate 26). The fleece-lined leather flying helmet was part of this flight suit, as were the fleece-lined leather gauntlets (plate 47). It was worn together with the fleece-lined flying boots. Rank insignia for all ranks from Unteroffizier to Generalfeldmarschall were stitched on to a cloth base of coloured material matching that of the suit. This insignia was worn in the same position as on the lightweight summer flying suit (plate 26).

3. *The two-piece fleece-lined leather flying suit for winter flights over sea areas.* This two-piece fleece-lined leather flying suit, was manufactured from dark brown or black leather (plate 27). It was worn together with the fleece-lined leather flying helmet, gauntlets and heavy duty flying boots. Rank insignia was stitched on to a base of thin dark brown or black leather matching that of the suit. Its position for display on this suit was the same as on the lightweight summer flying suit.

In addition to these three models of Flying Suits another form of flight clothing existed for use by Luftwaffe pilots and air crews:

The Electrically Heated Flying Suit. This was worn over the wearer's normal form of clothing such as the Fliegerbluse and Fliegerhose (flight blouse and trousers) and beneath the flying suit. Additional warmth was provided for the wearer by this suit when element leads from this suit were plugged into electrical points in the aircraft's interior. This suit had its limitations, the least of which was the restriction on movement that these plugged-in leads had on the wearer. The electrically heated flying suit proved an unpopular form of flight clothing and because of its high sensitivity was not in demand by air crews.

Luftwaffe Flight Jerkins: The use of flying jerkins or wind-cheaters was a form of dress that became very popular in the Luftwaffe, especially amongst fighter pilots. Movement inside the cockpit of German fighter planes was very restricted and a garment that was hard-wearing, close-fitting, warm and with zip fastened pockets was considered a more practical form of flight clothing than the issue one-piece flying suit.

By mid-1940 official permission had been given for these jerkins to be worn at the preference of the pilots and no doubt they helped to fill the widening gap left by the phasing out of the Fliegerbluse. As the majority of the jerkins were privately pur-

25. Above: *Four crew members of a Junkers Ju 88. All are wearing the summer flying suit with fleece-lined flying boots and inflatable lifejackets. From left to right: Oberleutnant, Leutnant, rank unidentified, Oberfahnrich (note the officer's-style Fliegermütze with silver piping worn by this NCO).*

26. Below left: *Contrast in Luftwaffe protective flight clothing. The Hauptmann on the left is wearing his parachute harness over the lightweight summer flying suit while the Hauptmann on the right is wearing the heavy weight fleece-lined 'Bulgarian' flying suit. The rank insignia worn on these two flying suits is identical in size and position.*

27. Below right: *The two-piece leather flying suit. The bear is the squadron mascot.*

24

Variations of Flight Jerkins.
28. Above left: *Oberleutnant Walter Oesau wearing the cream coloured cloth jerkin with knitted cuffs and welt.*
29. Above centre: *Oberleutnant Viktor Mölders (brother of Werner Mölders) wearing a form of leather flight jerkin.*
30. Above right: *Unnamed NCO wearing a leather jerkin with wool-knit cuffs and welt.*
31. Below left: *Knights Cross holder Hauptmann Nordman with a pilot officer. Both are wearing leather jerkins.*
32. Below centre: *A group of pilot officers wearing the blue-grey zip fronted cloth flight jerkins.*
33. Below right: *Flying boots were sometimes worn without protective flight clothing.*

chased they tended to vary in style and colouring. Some were manufactured from leather, some were fleece-lined whilst others were in hard wearing drill; and like the styles, the colours varied, from creamy white, through grey-blue to dark brown and black (plates 28-32).

Luftwaffe Footwear For Flight Personnel

The Luftwaffe provided its air crews with fleece-lined heavy duty flying boots (plates 13, 25). The foot and uppers were manufactured from black leather with the leg section to the boots constructed from soft black suede leather reinforced with black leather strips. The boots were worn by being pulled on to the foot then zipped up by the wearer. Straps and buckles, when fastened, helped to hold the boots on the wearer's feet. These boots were intended to be worn as part of the complete flying suit, especially in cold weather; they were, however, frequently worn by flying personnel without the flying suit (plate 33). During the summer months individual pilots with officer rank, often fighter pilots, sometimes wore their high boots together with officer's breeches, and the Fliegerbluse under their flying suits when on operational flights (plate 19).

Black leather lace-up shoes and boots worn together with long trousers were also worn.

D'

Flight headdress worn
by Luftwaffe air crews
during the Battle of Britain:
34. Above left: *The light-*
weight summer flying
helmet.
35. Above centre and 36, below
left: *The unlined light-*
weight tan coloured flying
helmet.
37. Below centre and 38, below
right: *The dark brown fleece-*
lined leather flying helmet.
This fleece lining is more
clearly visible in plate 38.
39. Above right: *In an effort*
to protect themselves
against gun-shot and splinter
head wounds, crew members
of bombers engaged on oper-
ational flights over enemy
territory, especially
through heavy flak, took to
wearing the issue pattern
steel helmet.

The Luftwaffe Flying Helmets

The Luftwaffe adopted three styles of flying helmets for wear by their air crew members. They were:
1. The lightweight summer flying helmet, constructed from dark brown netting and thin dark brown leather (plate 34). Its use was usually restricted to hot summer or tropical weather.
2. The unlined lightweight tan coloured flying helmet, which formed part of the complete lightweight summer flying suit. (See plates 35, 36, 44, 45, 48.)
3. The dark brown leather fleece-lined flying helmet. This was the form of helmet worn with both the heavyweight winter flying suit and the leather two-piece winter flying suit (see plates 24, 37, 38, 40, 46, 47.).

The Luftwaffe Steel Helmet: its use by Flight Crews:

Although the Luftwaffe steel helmet was intended for wear by personnel on ground duty, it was actually used by Luftwaffe air crews whilst flying over enemy territory. Experiments had been undertaken with various types of prototype protective head-dress designed to give crew members, and especially pilots, a degree of protection from gun-shot and splinter wounds. It was only towards the latter part of the war that a successful helmet was divised and actually put into production. In the meantime individual flight crews took it upon themselves to wear the issue pattern steel helmet whilst on operational flights (plate 39).

Luftwaffe Lifejackets

There were two patterns of lifejackets in use by Luftwaffe air crews:

1. An inflatable lifejacket was provided for fighter pilots and for airmen flying two-seater aircraft (plates 40-42).
2. A kapok-filled lifejacket was issued for wear by crews of bomber, transport and flying boats (plate 43).

Luftwaffe Goggles

Standard issue Luftwaffe goggles were manufactured by Leitz. They had fairly large interchangeable curved 'lenses' in plain glass or brown tinted glass (plates 43, 44, 47), the latter for use in bright sunlight or when flying over snow-covered terrain. Flying goggles of shatterproof glass were manufactured by Nitsche and Günther. Smaller than the Leitz model, they had smallish oval metal frames with either clear or dark tinted domed 'lenses' (plates 45, 46). The strap to these goggles was elasticated and adjustable.

40-43. Top row, left to right: *Luftwaffe lifejackets, 40 to 42 show the inflatable type, while 43 shows the Kapok filled variety.*

44. Below left: *The shatterproof flying goggles manufactured by the firm of Nitsche and Günther. (See also plate 2.)*

45. Below right: *The general issue Luftwaffe goggles manufactured by the firm of Leitz.(See also plates 43 and 46).*

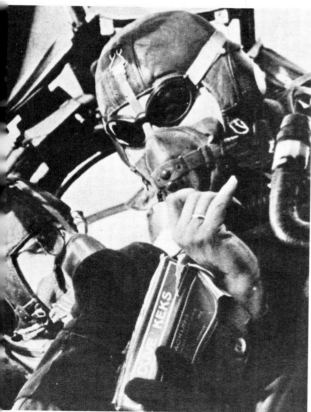

29

46.Left: *Bomber crew members wearing summer-weight flying suits and helmets with Leitz goggles.*

47.Above: *Oxygen masks for bomber crews.*
48.Below: *Luftwaffe parachute harness worn over the kapok-filled lifejacket.*

Luftwaffe Sleeping Bags

Quilted kapok-filled sleeping bags were provided for bomber and transport air crews. They were used on long flights by crew members who were not required to perform their particular duty until a given time (bomb aimers, cargo loaders etc.) Sleeping bags were carried in Luftwaffe air/sea rescue aircraft. They provided a form of extra warmth for rescued persons suffering from exposure. The sleeping bags were constructed large enough to allow one man to crawl into whilst wearing full flying uniform (plate 50).